CW00523298

Inoculate Your Biotech

Protecting and Boosting Your Product-to-Market Process

Kim Lim

Copyright © 2016 Kim Lim

All rights reserved. No part of this book may be used or reproduced in any manner whatsoever without prior written consent of the authors, except as provided by the United States of America copyright law.

Published by Best Seller Publishing®, Pasadena, CA

Best Seller Publishing® is a registered trademark

Printed in the United States of America.

ISBN-13: 978-1537703664

ISBN-10: 1537703668

This publication is designed to provide accurate and authoritative information with regard to the subject matter covered. It is sold with the understanding that the publisher is not engaged in rendering legal, accounting, or other professional advice. If legal advice or other expert assistance is required, the services of a competent professional should be sought. The opinions expressed by the authors in this book are not endorsed by Best Seller Publishing® and are the sole responsibility of the author rendering the opinion.

Most Best Seller Publishing® titles are available at special quantity discounts for bulk purchases for sales promotions, premiums, fundraising, and educational use. Special versions or book excerpts can also be created to fit specific needs.

For more information, please write:

Best Seller Publishing®

1346 Walnut Street, #205

Pasadena, CA 91106

or call 1(626) 765 9750

Toll Free: 1(844) 850-3500

Visit us online at: www.BestSellerPublishing.org

Table of Contents

Introduction

*"Hoping for the best, prepared for the worst,
and unsurprised by anything in between."*
- MAYA ANGELOU, I KNOW WHY THE CAGED BIRD SINGS

My main purpose in creating this book is to impart to you the idea that you don't have to dread weaving through the jungle of regulations and be scared off by years of putting a product through research and development to get a product market-ready. That there is a way to get your product to market as long as you don't get caught in the common pitfalls in the industry—all of the inefficiencies and regulations.

My intention is to prepare you for the worst and hope for the best. Being part of the process of making a product that helps save lives is much like having a baby. I forgot all about the pain and suffering of labor, when holding my son for the first time. I was washed over with love and serenity. I remember being filled with joy and anticipation, but then quickly scared to death and feeling overly responsible and protective. Here was someone who would depend on me forever – I mean *forever* – and I had no idea what I was doing. I still remember looking at him and thinking "Now what…?" The nurse must have read my mind because she looked and me and whispered, "You know more than you think you know."

It's the same for a product launch. When you are holding your first article off the manufacturing line, you can feel in awe at the power of a single medicine that will change someone's life and experience the butterflies in your stomach at the FDA inspection coming your way. But there is no way you got this far without gestating your idea, growing personnel, training programs, and outlining a path to this point. You know more than you think.

CHAPTER 1
Do Something About It

"Pearls don't lie on the seashore. If you want one,
you must dive for it."
— Chinese proverb

For years, I was an engineering consultant who traveled all over the US creating processes and building facilities for the pharmaceutical, medical device, and therapeutic industries. I remember I had to prepare myself for taking some Petri dish samples for environmental monitoring in a clean room area that was rated ISO Class 5. The gowning for this entails changing from street clothes into scrubs, a thorough washing of hands and donning a "bunny suit."

After spending an hour collecting the samples, I had to go to the warehouse and collect packaging to deliver them to the local laboratory for processing and testing. This took most of my day when I still needed to perform some other facility verifications. The next day, I got a call from the lab explaining that one of the samples was compromised during testing (they dropped it on the floor), and a replacement was necessary. I canceled my morning meeting and went back downstairs, changed again, scrubbed, and bunny-suited to collect ONE sample. After that, I went back to the warehouse to get another shipping container and had to drive that sample over to the lab for testing. This type of situation happened again and again.

Needless to say, I found I was having a hard time with the lab support services that were readily available. My goal was to move my client's products along in the process of getting approved by the FDA, and I was getting run into the ground with busy work and pointless tasks that were exhausting me and halting nearly all my forward progress. I started complaining to my colleagues about the lack of support I was getting and complained about it so much; someone finally asked: "Well, why don't you do something about it?" That's how I started Ultimate Labs.

With this book we're letting go of the typical and traditional methods and streamlining our efforts to get our products and devices market ready as quickly, easily, and efficiently as possible.

CHAPTER 2
Lesson Learned

"An ounce of prevention is worth a pound of cure."
– Benjamin Franklin

My first job in the biotech industry was working at a small medical device company that extruded catheters for feeding tubes and re-assembled molded parts into disposable scalpels. Here I was, home from college for the summer and figuring out what to do with my life. One of my tasks was to prepare for an upcoming regulated body inspection. I was given the company quality manual to review, improve, and revise. At the time, I had no idea how to use a Mac, let alone what a GMP was. The manufacturer and the job taught me many things, but the most important lesson was that sometimes you just have to work through the unknown. Working hands-on in the process, struggling to figure out answers can be the best way. No one can prepare you for answering questions from the FDA if you haven't read the quality manual or tried a process improvement. I remember being so proud that the manual was updated, and the calibration files were nicely filed, and stickers were on all the equipment. It didn't take long at all until I was embarrassed and disappointed when one of the non-conformances cited was a piece of equipment that "lacked proper documentation" and "needed further investigation." It was a microscope that was intermittently used and had a broken lens, and I had put a "released" quality control sticker on it. I learned that day that stickers are not just stickers; they are proof of what you did.

Preparing for an FDA inspection is much like going to the dentist. You know you need to do it, it's unpleasant, and it will reveal all the shortcomings of teeth-brushing and flossing from the previous six months. It's bad enough that you need to go for yourself, but running a quality department getting ready for an audit is like taking kids to the dentist. I have tried the gamut of bribes and threats to get them to brush their teeth every morning and evening. I've promised incentives, and I've had to wrestle them to the ground to get them to brush their teeth! Then it's the dreaded appointment. They are already scared and tense and then I hear the scraping away of plaque. I remember my son's first cavity and the ensuing crying and screaming during the drilling and filling of it. I promised myself then that whatever aggravation and fighting I would have to do on a daily basis would pale in comparison to the hand that was squeezing mine in pain and agony.

Thus, when it comes to all those batch records that need to be reviewed meticulously, do them slowly and carefully. Training sessions with personnel on GXP, repeat them over and over to the point that GXP is perfunctory. Cleaning the manufacturing floor for the hundredth time in a week? Do it again for one hundred and one. All these little actions that need to be done on a daily basis are extremely important when it comes to the day of the inspection. It is the worst feeling to open a cleaning log and the page has blanks. If you can't prove you clean on a regular basis, how can you prove that you have due diligence in your manufacturing process?

CHAPTER 3

Creating Your Quality Process Blueprint

The best laid schemes o' Mice an' Men,
Gang aft agley,
An' lea'e us nought but grief an' pain,
For promis'd joy!
– ROBERT BURNS

You have a great new medical device that will change the world, or maybe you're just trying to make a better version of a diagnostic test that's currently on the market. Now what? My recommendation is to create a blueprint of your process to give it structure and an outline on which to build your people and your systems. Creating one provides an overall picture of your product's future. Be aware that your blueprint can change and can be revised as many times as you like. Things to consider when you're defining your blueprint:

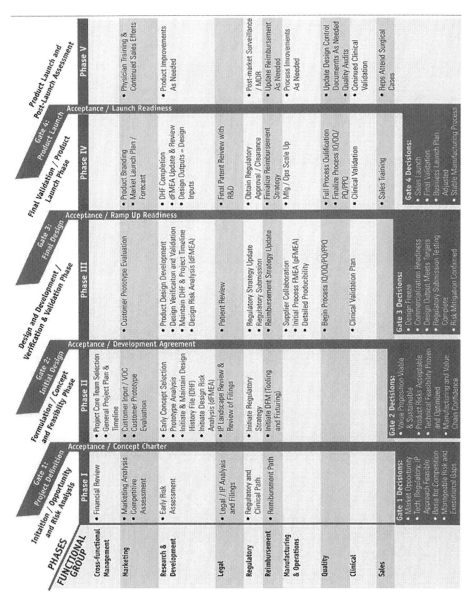

From: http://medicaldesign.com/technologies/model-device-development

There are two sets of initial questions that you need to answer when defining your blueprint. First, you have to **define your product.** Here are things to consider:

Product Description

- If your product could talk, how would it explain itself?
- How does it work and what is the science behind it?

Product Phase

- If your product was a human, what stage is it in?
- Is it a baby who needs lots of care and attention?
- Is it a child ready to go to school and socialize with other children?
- Is it an adolescent who is under a lot of peer pressure and market scrutiny?
- Or is it an adult, whose lessons are all learned and only modifications need to be made?

Product Impact

- Does your product fill a need or a want?
- Do you really think the liquid form of a male enhancement drug is on the market and can it compete with a treatment for a Non-Hodgkin's lymphoma?
- Are you working on the next HIV vaccine, or are you working on the next over the counter pain-reliever?

Secondly, you have to consider **your company history.** Are you an engineer working on a new medical device and you just have this great idea but no business sense? Are you a scientist or a researcher who has found something new and has NIH backing, or grant backing, or have any kind of institution that can help you promote your discovery? Do you have no technical experience and you're just a business person looking for a way to break into the bio-tech industry?

Another thing to consider is your company culture and your business risk. What kind of person are you and what are you willing to do? Are you an entrepreneur and willing to take lots of risk? Are you conservative and need venture capital funding, structures in place, and lots of support systems? What size risk are you willing to take with your product?

Here are a few examples where these types of blueprints are overlaid in small-business, medium-sized-business, and large-business scenarios.

Small Business

For a small business, you are bootstrapping and burning the candle at both ends. In addition to the science you are trying to prove, there are so many things to worry about along the way, such as sales, marketing, finance, insurance, employees, and infrastructure.

Here is my best advice in this scenario. Shop, shop, and shop some more – outsource as much as you can. You don't want to be worrying about payroll was when you're looking down a microscope at 1 o'clock in the morning trying to figure out why your cells are not undergoing lysis correctly. When times are tough and bills were racking up, business becomes very personal. The bank that was so willing to help you and supportive of your venture is now knocking at your door wondering if you're still in business. That failed marketing campaign that cost you thousands of dollars is now subject to a lawsuit. Many investments of money or time become costly mistakes.

It's true when they say that business is business, but to a small business owner when bankruptcy is looming over you and there's a threat of your home is being taken away, it can be a huge distraction from that goal you've been striving for – which is saving a life or making one better.

My second best piece of advice is don't stop trying. There will be lots of mistakes. You're going to learn a lot, but you're still going to get a product to market. There was a small business that used to rent a little bench in my lab and they grew into a business designing small molecule drugs.

Day after day, I'd watch them persevere through projects and personnel changes with aspirations of expanding and they finally did. Not in a month or two, or even three, but after a year and a half. It took them lots of perseverance, patience, and persistence. That is the scenario for a small business, and it's not for the meek-hearted – it is however, very rewarding.

Medium-Sized Business

Now you have a great staff in place, good systems and a fair understanding of what a regulatory body expects of you. There are learned people who are in quality assurance and in regulatory affairs who can walk you through the different phases and the different designs of what it will take to get your product to market. You have a clear understanding of what the market analysis is, you have a quality plan, and a clinical path. Perhaps you have even gone so far as to get a market and sales forecast. All of your feasibility studies have been done at this point and your engineering specifications are in place. You are trying to grow your business and to make sure that you can execute the projected business plan. The little things that you don't realize are important are the details of design, changing your design and having good systems within which you can maneuver and make these changes. Being able to be flexible in the decisions that you're making is key. An example of a company that's a medium-sized business is a client that I have watched grow from just 12 people to 500 people on the platform of doing DNA sequencing and genomics. With the technology that has changed, there's a need in the market for it and since they have used a good-quality plan, they have grown into a large company.

Large Business

In a large business scenario, history propels your business forward. There are many products on the market, and there are even more in pre-market status that have the science and technology, but some companies do not have the big business structure to launch their product. This is when an experienced CEO and respected COOs in the bio-tech community seek out board opportunities to give an organization good guidance and help them make profits. One of my past clients had a great business plan for manufacturing generics from patent expiring blockbuster drugs. Imagine a pharmaceutical like Gentamicin, whose patent, used for most antibiotics, was coming close to expiration. The company decided that they wanted to make a generic version of it and sell it on the market. It was a great idea.

Their ANDA was approved, they had built a multi-million dollar facility and a manufacturing process was in place. Unfortunately, somewhere along the way, the development department didn't create tests to prove consistency of the potency of the drugs that they were making. They were slapped with a consent decree and had to validate all the processes and systems that they had in place. Part of that process was to perform a cleaning validation, where one verifies that there is no cross-contamination of a cleaning agent or other formulations in the processing vessels used for manufacturing. My job at the time was to be lowered into several 10,000-gallon tanks to determine whether the cleaning spray pattern inside was enough to remove all the residue from the last batch made. Even with these horrifying brushes with worker's compensation claims, the company couldn't recover and finally went bankrupt. It was a great idea, but it was poorly executed.

CHAPTER 4
Practical Quality: The Risk-Based Mindset

"Risk comes from not knowing what you're doing."
– WARREN BUFFET

We've all been there: How many times have you heard, "Where I worked previously this is how we did things…"? Quality systems are sometimes a montage of the different perspectives and experiences of several people. Not only that, but then we end up adding to, revising and repeating without thinking about where this concept came from or how this came about. How many times have you seen a 26 days' work instruction or a cluster of decision sheets that try to account for every exception to the rule? Why does this happen? This happens when we lose sight of the most important questions: *How do we impact our product?* and *How does this process impact the product that I'm making?*

An Approach for a Risk-Based Mindset

Our industry is riddled with seminars and white paper about risk-based approaches. I challenge you to adopt a risk-based mindset. "Approach" implies that that we do a certain process temporarily to fit the situation. A mindset is a way of successfully determining the best course of action in any case. It's the difference between spending time, money and effort to adhere to a system or spending that time, money and effort to ***improve*** a system.

What is one of the worst meetings you will ever encounter in this industry? An audit. It doesn't matter if it's a regulated body, a potential customer, or your internal audit – they are unnecessary and painful most of the time. The purpose of these meetings is to dig deep into your systems, put your staff on the defensive, and make you squirm in your seat waiting to hear if the corrective actions for your facility or product are complete and ready for manufacturing.

This may be difficult, but try to imagine what it would be like to be an auditor. You've been tasked by a government agency or quality department to determine whether a company is following the rules and regulations that protect public safety. There's a gate between discovery and progress, and you're the gatekeeper. The weight of this responsibility is heavy, mostly under-appreciated, and underrated. The things that you sign are the law, so you have to take your job seriously. Now that you understand the other side of the audit, I have two absolutes to help us adopt a risk-based mindset in order to help get us through audits, company growth, and different scenarios.

Justification is needed. As long as you can justify why you chose to do something in a certain manner, it's generally accepted. After that's attained, then whatever procedure you write – a batch record, device history record, or a design of experiments – you need to follow through. Do what you say you are going to do. And here's a little bonus for you: learn from your previous experience. Be resilient and be open-minded. These are key components to having a risk-based mindset.

Justification

Justifications must be supported by sound risk analysis and logical thinking. There are excuses for doing things; those are not justifications. See if you can tell which is which: a justification or an excuse.

Case Study A

Scenario: A contract manufacturer started making small batches of cell matrices. They had to perform sterility testing per a USP guideline and cell into the small batch category. The problem: They only made about ten 1mL vials per batch and they needed all the material not only for the battery of tests, but to actually put on stability for months at a time so they can tell if the potency was going to remain intact over a period of time.

The standard states that sterility testing has to be done on three separate lots, and a percentage of the lot according to volume or number of vials. The manufacturer decided to test one vial as it was 10% of the whole lot. Is this a justification or an excuse?

Justification: There is statistical analysis that was rendered during the discussions for the lot size they were going to use for testing sterility. It was a well thought-out risk analysis that maintained the quality of their product and was cost effective for the stage that they were in: phase 1 and phase 2 testing of their product to bring into market.

Case Study B

Scenario: A manufacturer has a clean steam system that needs to be tested at a regular interval for bacterial and endotoxin testing. The clean steam system typically has steam traps that condense the steam into liquid for testing. It was determined that this needed to be done on a monthly basis considering that they only manufacture twice a month. They were willing to risk that within a month, the clean steam would not be compromised or compromise public product safety.

There's also a rule that dictates the sample size of the testing. It states that 100mL of the clean steam condensate must be used for testing. The company decided to use only 60mL because that was all they could collect in one shift during the day. Justification or excuse?

Excuse: There are definitely ways that you can get a sample. The standard is very clear about having a full 100mL sample because it is a highly purified type of water that cleans steam condensate. It needs to be tested accordingly. The standard must be followed in this case.

Case Study C

Scenario: A medical device company was expanding their operation and adding more clean rooms to their manufacturing area. As part of this, they needed to do an FDA submission addendum to prove that the pre-construction process in manufacturing had not been compromised or modified. They also needed to prove the modified facility could produce the same quality product. The original testing of the facility included verification of the air handling system and environmental monitoring of the area.

The addendum testing was required, but could be scaled to just the modification. There were several discussions considering resources, data analysis, what the outcome of the data would be, and the manufacturing schedules that were already in place. Whether the entire facility or just the modification needed testing was the question. It was decided that the entire facility was going to be tested. Again, is this a justification or an excuse?

Excuse: They decided that the addendum testing was required, and it could be potentially scaled for just the modification of the clean room. They decided to do the entire facility and retest and did excessive testing. Again, this is kind of a case where it is an excuse. There is absolutely nothing wrong with the direction that they chose to sample the entire facility, yet again, it's just excessive work to explain to an auditor or even the FDA submission. To just test that this particular area meets the same standards as the original facility is sufficient enough to show that they are in compliance. They have the facility under quality control. Is this just an excuse to do everything, so no further risk analysis is performed? Some thought this process used would lead to more cost-effective testing.

Case Study D

Scenario: A pharmaceutical company that had a vivarium was cited for inadequate testing of their facility at a regular interval. The response to the citation was an investment in a $450,000 piece of equipment that was used to process testing samples more efficiently without having someone going to collect samples. After several failed attempts to verify that the equipment functioned, it was archived in the basement. All the actions that were associated with purchasing the equipment, trying to install the equipment and testing it, were logged into the quality system and it was cited to have resolved all the inadequate testing of the facility. But did it really? Is this a justification or just another excuse?

Excuse: There was a response to the situation of the investment in a piece of equipment that was processing samples more efficiently. Did it really though? It ended up being a large paperweight archived in the basement. This solution was logged into how it resolved – the inadequate testing of the facility. I believe this is an excuse because it wasn't truly addressing the issue at hand. There was no root cause analysis to show why they were not doing testing at regular intervals. They tried to implement an automated testing solution that didn't work. They tried to justify why using human resources wasn't an option for this and all they had was an excuse. In an audit situation, this would not be acceptable.

These are kinds of the examples and things that I have seen in my experience. Do we really understand how our product is impacted? Can we do risk-based, logical thinking? Have a big picture in mind of what you are trying to accomplish. Does what you're doing impact the product process or the safety of your product?

Follow-Through

The second absolute that you need for having a risk-based mindset is follow-through. Follow-through is simply doing what you say you are going to do, or say what you're going to do and do it.

Case Study E

A small medical device manufacturer wanted to improve their packing process. There were several engineers who toiled over the actuator speed and the vision inspection system. There were lots of interdepartmental meetings discussing the location and the footprint of the machine. Facilities laid out all of the utilities that supported the machine, such as compressed air, electrical systems, etc. Several people had different tasks to support the endeavor, and it was close to a $600,000 budget to get this packing machine into the facility. There were permits that were executed according to the drawings and specifications, so the city was involved in making sure this met city code specifications.

The delivery and installation of the equipment was followed by a fire marshal inspection. It was found that the end of a conveyor belt line exceeded the minimum personnel clearance by two feet. After a ton of finger pointing (and digging), a facilities technician admitted that he had eye-balled the measurement of the room instead of getting a measuring tape to verify it exactly. Lack of follow-through cost another $300,000 of modifications to the equipment to accommodate the actual room size. Not only was this money intensive, but it was also time intensive. They lost out on their aggressive timeline to bring their product to market.

Case Study F

A medium-sized business was going through an audit and the auditor noticed that there were some pest issues. There are the rats that run around facilities and clean rooms, but bugs and insects need to be under control. One of the several citations that I have seen in the past is "pest control logs are not up-kept well." The auditor in this situation asked for the pest control log and there were daily entries of people examining the facilities and logging in that there were no pests around. Obviously, there were several insects running around the facility.

Again, when you sign something you have to be accountable for it. You have to do what you are signing your name next to saying you'll do. If you do not, it is a violation and you can get a citation. There are lots of implications that come along with that. In this particular case, the facility was shut down for a period of time to get the pest situation under control. Having the log compromised has been an issue that the FDA has cited over and over, and it becomes public record. Anyone looking at this company can now see that they had issues with pest control.

Case Study G

A large business had some sterility failures on some of the lots of the pharmaceuticals that they were making. This company was making hundreds of thousands of vials of a generic medication, and they sent their particular sample size to a laboratory for testing in order to secure a release. A part of that testing is sterility testing. It takes fourteen days for this test to follow-through, so there is already a stop of promoting and marketing. Timelines are really important and materials are expensive. Personnel time is important as well.

In this particular case, the sterility failures kept happening over and over. Each batch was worth about $1.5 million. Having that on hold for a long period of time cost large businesses lots of money. There was a root cause analysis done and it was found that the autoclave was not working properly when they were doing a terminal sterilization on all of these vials. There is a printout that goes along with the autoclave that shows you how high the temperature has gotten for a period of time as well as the pressure of the autoclave to make sure that your cycle is actually killing off any microbials in the vials during the time of autoclaving.

In all of the logs for the autoclave, there are printouts. They are obvious. They have temperatures on them. They have lots of data, and someone is tasked to review that data and log it into a system to say that it's okay. This particular person just happened to be on autopilot and thought that all the lots were going to have the same outcome on the printouts for the temperatures, and although there was one instance where the temperature did not meet the acceptance criteria, it was signed off anyway, and now the lot was in question. That is a $1.5 million mistake on follow-through.

CHAPTER 5

Practical Quality: Compliance and Cost-Effectiveness

"Efficiency is doing things right; effectiveness is doing the right things."
– Peter Drucker

Important factors for practical quality are compliance and cost effectiveness.

For a medium-sized business, there was a regular sampling schedule and, over time, some of the sampling resulted in failures. Through a new Head of Quality, the failures were not being addressed. There were a lot of suggestions about how to do a root cause analysis in a facility, examination, and investigation.

Determining this and following through with a root-cause analysis would show that there was a facility flaw. What did they do? Instead of addressing the issue, they decided that the testing was the problem and, later on, there was an audit. The auditor found that there were other indications that there was a facility flaw.

It could have been easily fixed, and the citation could have been avoided. As a result, their production processes have been delayed, and they also have had to implement more rigorous testing policies at more frequent intervals that are of course, more expensive.

Another example is of a large business that had gone to market with their product. They had a sequencing diagnostic kit that was on the market already. After a review of the clinical studies, it was found that someone had not reviewed the data thoroughly enough. It was submitted anyway and the productions for the business had already been submitted to their board of directors. There was a lot of hiring and five hundred people were hired to produce all of the kits.

In the end, the clinical board denied their submission and the data was considered falsified because it was not reviewed correctly. That resulted in laying of 30% of their workforce and marginally putting this product out on the market. Ouch! If any of this makes your heart drop, go back and re-read the end of the last chapter where I summarize the enormous importance of follow-through. It could save your company millions…and even your co-workers' jobs!

CHAPTER 6

Hidden Costs and Unexpected Problems

"Why do we do the things we do? Save lives through testing."
– Ultimate Labs

We have talked about several different things throughout this book: how to decipher regulations (with a few case studies about that), how to create your processed blueprint, and following a blueprint to understand how quality systems work and how you can implement them. Then we talked about practical quality and having a risk-based mindset for your systems and methods. We also talked about the idea of doing it right the first time, and taking a look at where you are in your process development as well as performing things that are appropriate during that time and for your budget.

In this chapter, we're going to talk about a universal case study and how to determine hidden costs and unexpected hiccups in your process that you may not see right away. This will show you how to derive solutions and implement of all of the skills that we have just learned.

This example applies to all companies: those that make therapeutics, biologics, pharmaceuticals, even medical devices – anything that needs a water system to support product production; cleaning the facility, or anything that water touches at all.

How to Create a Compliant Water Testing Program

Why should we talk about water testing? Whether you know it or not, it is an essential part of your process. You have a great product. Your company has great aspirations for your business, and you realize that you need this utility system of purified water to help support your product. Whether it's directly in your product, a feed source to your product, or even used in cleaning your facility or the equipment that your product is being made on, it is an essential part of your process. Chances are you are spending way too much time and energy to maintain compliance for this water.

To figure out where to start, let's talk about where the regulations on the water come from. The USP has a distinct regulation for purified water, and if you've ever seen it, you know how daunting it looks.

Purified Water

[NOTE—For microbiological guidance, see general information chapter *Water for Pharmaceutical Purposes* ⟨1231⟩.]

H_2O 18.02

DEFINITION

Purified Water is water obtained by a suitable process. It is prepared from water complying with the U. S. Environmental Protection Agency National Primary Drinking Water Regulations or with the drinking water regulations of the European Union or of Japan, or with the World Health Organization's Guidelines for Drinking Water Quality. It contains no added substance.

[NOTE—Purified Water whether it is available in bulk or packaged forms, is intended for use as an ingredient of official preparations and in tests and assays unless otherwise specified (see *8.230. Water* under *8. Terms and Definitions* in the *General Notices and Requirements*). Where used for sterile dosage forms, other than for parenteral administration, process the article to meet the requirements under *Sterility Tests* ⟨71⟩, or first render the Purified Water sterile and thereafter protect it from microbial contamination. Do not use Purified Water in preparations intended for parenteral administration. For such purposes use Water for Injection, Bacteriostatic Water for Injection, or Sterile Water for Injection. In addition to the *Specific Tests*, Purified Water that is packaged for commercial use elsewhere meets the additional requirements for *Packaging and Storage* and *Labeling* as indicated under *Additional Requirements*.]

SPECIFIC TESTS

[NOTE—Required for bulk and packaged forms of *Purified Water*]
• **TOTAL ORGANIC CARBON** ⟨643⟩: Meets the requirements
• **WATER CONDUCTIVITY,** *Bulk Water* ⟨645⟩: Meets the requirements

ADDITIONAL REQUIREMENTS

[NOTE—Required for packaged forms of Purified Water]
• **PACKAGING AND STORAGE:** Where packaged, preserve in unreactive storage containers that are designed to prevent microbial entry.
• **LABELING:** Where packaged, label it to indicate the method of preparation and that it is not intended for parenteral administration.
• **USP REFERENCE STANDARDS** ⟨11⟩
    USP 1,4-Benzoquinone RS
    USP Sucrose RS

USP Monographs

Your purified water is the lifeblood of the entire operation. There are many people up and down your organization who will be concerned about it throughout manufacturing, quality, even in the facilities department who have to maintain a water system like this. It touches quite a few departments in your cells. Part of the pharmacopeia tells you that you need to do certain tests for this water. When do you do these tests? How many times do you do these tests? Can you afford to do all of these tests? Here are some of the most important things about your water that we're going to dissect out of the regulations. We will learn how you can apply what the regulations are telling you – how to interpret it with a risk-based mindset – and understand where you are in your process.

If you are at the beginning of the development of your product and you have already implemented a water system – even if you are buying water and utilizing water through a third party – there still needs to be testing to make sure that you are compliant with the regulations and you have built some safety measures into your product. The pharmacopeia has what we call a monograph in there. A monograph is just a fancy way of saying, "This is how you make this water and this is how you know you made it properly."

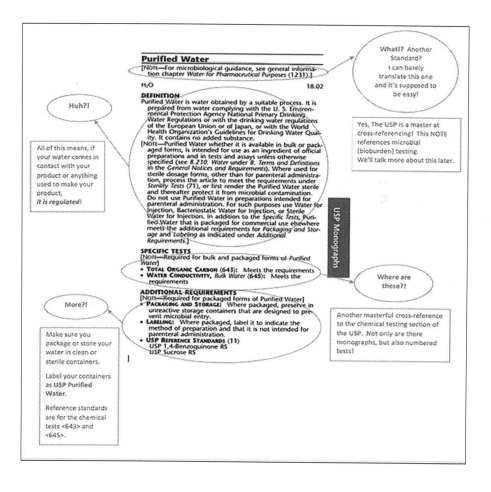

Purified Water

[NOTE—For microbiological guidance, see general information chapter *Water for Pharmaceutical Purposes* (1231).]

H₂O   18.02

DEFINITION

Purified Water is water obtained by a suitable process. It is prepared from water complying with the U. S. Environmental Protection Agency National Primary Drinking Water Regulations or with the drinking water regulations of the European Union or of Japan, or with the World Health Organization's Guidelines for Drinking Water Quality. It contains no added substance.

[NOTE—Purified Water whether it is available in bulk or packaged forms, is intended for use as an ingredient of official preparations and in tests and assays unless otherwise specified (see *8.230. Water* under *8. Terms and Definitions* in the *General Notices and Requirements*). Where used for sterile dosage forms, other than for parenteral administration, process the article to meet the requirements under *Sterility Tests* (71), or first render the Purified Water sterile and thereafter protect it from microbial contamination. Do not use Purified Water in preparations intended for parenteral administration. For such purposes use Water for Injection, Bacteriostatic Water for Injection, or Sterile Water for Injection. In addition to the *Specific Tests*, Purified Water that is packaged for commercial use elsewhere meets the additional requirements for *Packaging and Storage* and *Labeling* as indicated under *Additional Requirements*.]

SPECIFIC TESTS

[NOTE—Required for bulk and packaged forms of Purified Water]
• **TOTAL ORGANIC CARBON** (643): Meets the requirements
• **WATER CONDUCTIVITY,** *Bulk Water* (645): Meets the requirements

ADDITIONAL REQUIREMENTS

[NOTE—Required for packaged forms of Purified Water]
• **PACKAGING AND STORAGE:** Where packaged, preserve in unreactive storage containers that are designed to prevent microbial entry.
• **LABELING:** Where packaged, label it to indicate the method of preparation and that it is not intended for parenteral administration.
• **USP REFERENCE STANDARDS** (11)
  USP 1,4-Benzoquinone RS
  USP Sucrose RS

USP Monographs

What!? Another Standard? I can barely translate this one and it's supposed to be easy!

Huh?!

All of this means, if your water comes in contact with your product or anything used to make your product, *it is regulated!*

Yes, The USP is a master at cross-referencing! This NOTE references microbial (bioburden) testing. We'll talk more about this later.

Where are these?!

More?!

Make sure you package or store your water in clean or sterile containers.

Label your containers as USP Purified Water.

Reference standards are for the chemical tests <643> and <645>.

Another masterful cross-reference to the chemical testing section of the USP. Not only are there monographs, but also numbered tests!

If we take a look at it, we can break it down piece by piece.

In a nutshell, this particular monograph is telling you that there are three separate tests that need to be performed on your water, and it verifies that it meets the standards for purified water. Two of the tests are really obvious: the total organic carbons and the water conductivity. It's stated right in the middle of the page. But there is another hidden test, the third test that has some subjective views about whether it should be done when it should be done, and how it should be done. The USP is notorious for hiding things within other things within a regulation. If you read the definition section of this monograph, you'll get the clue for the third test, which is microbial contamination and how to do fire-burning testing.

Because purified water is the lifeblood of your facility and your products, we need to understand what it means to do these three different types of tests, and that we need to do them in a way that promotes an understanding of how they're performed, who should perform them, and some of the pitfalls of the test. You can adjust to make sure that the water is compliant for the stage of your product development process that you're at.

Validation

At this juncture, there is a scary word you will hear: validation. Validation is a term meaning to verify that your system is producing consistent, regular, purified water on an ongoing basis. Let's talk about what that really means in your stage of the process. You've just created this system, or you're buying this water, so you need to validate or verify that this water is going to meet the standards of the product that you're making.

A system qualification is done, so if you have the system in-house, you check that all the components of the system have been installed correctly, that they're operating correctly, and that they're going to perform correctly over a period of time. You want to see reproducibility and repeatability from the system creating water. This is a whole industry within itself. This requires some engineering background and an understanding of how water systems work, as well as a good understanding of how you're going to maintain it over a regular period of time.

During the performance qualification, we can adopt a risk-based mindset and understand that depending on your location, and how often you will be manufacturing, we can decide on what a performance qualification sampling plan looks like. If you're living in San Diego or southern California where the temperature is moderately temperate, then a 30-day cycle – which would be representative of most of the climate during a year – would be enough to show that you have control over your system. If you're living in the northeast where the seasons change on a regular basis, it may make more sense for you to do a 30-day cycle once a quarter to understand your water system and how it's behaving during those periods of time. In the winter months, if you're dropping to a really low temperature close to freezing, then your conductivity and TOC readings are not indicative of what is going on in your system. In the summer months, when it's hotter and there may be more flora in the air, you're going to have more contamination or a biofilm in your water system that you would need to know about and would want to monitor on an ongoing basis.

There are really distinct ways that you can justify why you're doing the testing that you're doing, and statistical analysis or a quantitative plan to justify why you are doing what you are doing to test your water system. This goes back, again, to how much you're going to follow through on what you say you're going to do. We talked about that before in justification and follow-through.

Now, we also need to take into consideration how often you're going to be manufacturing, and how often you're going to use this water system. If you're using this water system day after day, then it makes sense that you're going to do weekly testing or bi-weekly testing of your water system. Remember that the risks that you're taking are from the last test to the next test. If you have some kind of contamination, are you willing to risk the entire product that was made during that period of time, or are you willing to risk whatever process your product has been exposed to at that particular time? Is that acceptable to you?

Those are the kinds of questions that need to be asked at this juncture. For example, if you are just using your water system for cleaning the facility and you're doing ongoing monitoring of your facility and on Monday, your water system looked like the test results were fine, and then the next Monday they don't look so great, do you have the opportunity to fix the system, and to allocate resources, people time, and materials to make up for that lost week of production? Do you need to clean the facility over again, on top of trying to troubleshoot why there was an apparition in the testing to begin with? Those are things that need to be considered in that situation (in which you have a risk-based mindset).

Let's say you're only manufacturing twice a year, and you have done your performance qualification on your system and it seems to be in compliance and control. Now, with a risk-based mindset, you may look at it as, "Okay, I'm only manufacturing in the months of April and November. I'm going to do water testing weekly during the months of April and November, and move to a testing scheme of once per month or every other month in between." That is justifiable and it makes sense, and as long as you are in compliance, then you have no problem.

I have seen other facilities where they have ongoing weekly testing no matter what, even if they are manufacturing for just two months of the year, seemingly to keep up with ongoing testing requirements. However, is it really necessary? If you can show that your system has had control in the past, then it's justifiable to reduce the testing criteria that you have and reduce the frequency of the testing.

Let's swing the other way now. Let's say you do have a problem when you do your testing in April and that it has an issue that you have to resolve. Then you need to show that you're reproducible and repeatable in making the correct, standard water. There may be an increase in testing for at least a few months until the next time when you're doing manufacturing to prove that you are still in control. That's what the FDA is looking for, and so are your clients and your customers. They're just looking for proof that you are in control of your process and your system.

What Water Testing Does For You

Having conductivity readings gives you a good indication of any inorganic materials that you may have in your system, which also gives an indication of where your system is at maintenance-wise. If you have a good preventative maintenance program and you are changing out your membranes and flushing out your lines on a regular basis, changing your resin beds and your ultraviolet lamps, then your system should be in compliance. A conductivity test is a really good indicator that some of these things are not in compliance, and it's an easy test that can be done on a regular basis with regular monitoring.

For a total organic carbon, it's a little trickier. It shows organic contamination and biofilm that may have wormed its way within your system and in your lines. This may require a good cleaning and sanitation of your system. These tests – conductivity and TOC – are good indicators that your system is still in control. If it is out of control, it's an easy fix at this point in time to make. You can do some maintenance on your system, or you can do a system sanitization in-flush, which is low-cost at this point.

A heftier test is the bio-burn test, which shows how many organisms you may have in your water system, or it's an indication of how your system is behaving over time. The regulations are subjective and a bit tricky; it shows that you don't have to do this test, but it's recommended that you do it. It doesn't give you any distinct criteria at all. It has a guideline to show you what the industry standard is, but it really encourages each individual to determine what their specific criteria is and create a baseline test for their bio-burn and try to adhere to what the baseline was during a performance qualification or historical data.

There are several ways to do bioburden testing. The most accurate way is to perform membrane filtration. It's accepted in the industry and most of the time, if you find bacteria in your system, it can be traced back to the feed-source water. That is easily corrected by changing filters out, adding more filters to the system, changing out your ultraviolet light lamp to kill off more of these bacteria, or even changing out some of the membranes in the system that may have been overused or overworked.

These are typical ways to avoid the pitfalls of water systems. Keeping these factors in mind, you can avoid both over- and under-doing your water testing, be in control of your system, be in control of your facility, and ultimately, be in control of your product with a high level of assurance that you are making the best product that you can make.

Hidden Costs of Doing Water Testing By Yourself

Do you have the capital to dedicate personnel to water testing on an ongoing basis? How about the facility? How about the equipment, the capital equipment expense, and the ongoing maintenance expense of having all of this specialized equipment to do these tests on site? Lastly, do you even have the space for any of this, the time, or the technical acumen to look at these things, interpret the results, and apply these troubleshooting ideas and mechanisms to make sure that you're still in compliance?

This illustration is to show what it typically costs to do the testing yourself versus what it costs to do it through a contract laboratory service. I hope this helps you decide what is cost-effective for you, and how much time you have to dedicate to something that is a support utility to your main product.

Overstocking inventory
with lab supplies

Running complicated test instruments
with multiple people hours

Hours of of review cycles while de-
ciphering complicated data

CHAPTER 7
Big-Picture Questions

"The first bowl of chocolate pudding was too hot, but Goldilocks ate it all anyway because, hey, it's chocolate pudding, right?"
- Mo Willems, Goldilocks and the Three Dinosaurs

In this chapter, we're going to walk through the process of discovering some of the questions that you may have and illustrate what it means to implement a risk-based mindset with justifications and follow-throughs. Think of this chapter as the Goldilocks test: Are you too hot, too cold, or just right?

My first example is about a medical device company who had an interesting excursion happen. They had a fire over a weekend and I was called to assess what the damage was. They had a pretty good manufacturing facility with processes in place. They were FDA approved, ISO certified, had good systems in place, and obviously a good emergency response system. The fire department came immediately and put out the fire.

However, that caused a whole facility shutdown because the entire facility was flooded. In a clean room environment, it's very difficult to remediate. The walls needed to be pulled out. They needed to be dried and then a new sealant needed to be put in to make sure that it could meet the environmental conditions and constraints that they had in the past.

Given that they had to move most of their processes to a different area, what came into question was: Was there enough space to actually execute their usual, ongoing production process, or did they need to come up with a temporary fix until the new facility was up and running?" The big question when they were assessing the risk, finances, production, and manufacturing was: Could they stay shut down for the period of time that they needed to?

To answer that question, there were several smaller questions to answer first and many other factors to consider. In a perfect world, the shutdown may last two to three weeks, so what was the cost of the plan to create a temporary situation? Given the size of the company, their production in the past did not exceed their sales forecast. Their sales department was doing an excellent job of selling their product on an ongoing basis. The quotas that they were making every quarter were not even meeting the expectations of the sale. Now they were back ordered. That posed a problem.

There was a sense of urgency; they needed to figure out a solution to what was going on, or put in the timeline that there was going to be further delay in product delivery. Was there an option to size down some of the ordering? Doubtful. It was considered a new product on the market, well-vested. Also, it had been accepted by the insurance companies – a huge hurdle for most medical devices and pharmaceuticals. To lose traction in the market would have been devastating to future forecasts.

They decided to see if they could construct some of the other parts of the facility to accommodate the manufacturing. There were several things that they considered. Was this going to be in compliance with their FDA filings and their ISO filings? Did they have enough procedures, policies in place, and standard operating procedures to accommodate this kind of move and to accommodate any manufacturing modifications that they would need to make?

Then there was, of course, the timeline. It was urgent. Would this be an ongoing situation that they would upkeep, or was it just a temporary fix? How long was this temporary fix going to last? They had to consider all the different types of utilities that they were using. Not only do you have to consider the environment, i.e. the clean room classification, but also all the supporting utilities, water, compressed air, any other compressed gasses like argon or nitrogen. All of those had to now be piped into the area where they needed a temporary fix.

After the timeline, we began monitoring the environment and seeing if we could meet the clean room classification that they had in previous manufacturing. We also looked at some of their past trends; if they had some problems, was this going to reoccur again? If they had excursions where they had high levels of bacterial counts, or high levels of particulate counts in a particular area, or a specific production process, was this going to translate into something even worse in a temporary environment?

Finally, they came to a decision that this situation was going to work out for them for at least three months. They constructed the temporary facility, brought it up to speed within a week, which required people working twenty-four hours a day, seven days a week to get it up and running. They had to expend every effort to move production and process equipment over, install and test all of the equipment before a full manufacturing came up and then implement all of the training programs for their personnel. It cost $600,000 to do this. In the end, it paid off. The products that they were making during that particular time grossed about $4 million worth.

The gain was definitely worth the pain of the two-to-three weeks when the transition happened. In just over two months, the facility was brought back to its original state. Things moved yet again. There was a lot of talk in finance about how it would be more expensive to move back to the previous facility, but given the numbers of the manufacturing and production, and given the forecasts, they decided to spend another $400,000 to move equipment and people back into the original facility and make the temporary facility a permanent situation. They ended up grossing close to $8 million worth of product in a given quarter and moving their process forward.

That's an example of doing some good risk-based mindset decisions – doing some justification along the way – and also, following through on what you say you're going to do. Getting quality working with manufacturing, manufacturing working with finance, finance working with regulatory and having a clear plan of attack to make things go as efficiently and quickly as possible is key for you to maximize the rate of your return. That is an example of the entire process.

Let's talk about something a little less grandiose. An everyday, day-to-day operation and what a simple decision can do to create the larger picture in an organization.

There was a company that had a therapeutic. They were filing for their phase 3 production and phase 3 clinical trials and needed some initial data on how their product was performing. Particularly, they were trying to determine their endotoxin levels and their bioburden levels of what their product had been producing in the past.

The process was very labor intensive and was an aseptic process, which means that it's sterile all the way through and uses very small production batches because it was very expensive and had a cellular matrix to it. It was very, very difficult to manipulate without having human contact, or some laboratory manipulation to go along with it as well.

When we received the final product, we tested for endotoxin and tested for bioburden. There were questions about what levels were acceptable and how to come up with a justification of the acceptable levels. This was a very interesting case because it depends on how you look at what it is.

If it is a cellular matrix and the application is somewhat cosmetic, does that mean that it's a cosmetic, which falls under one set of regulations? Does it mean it's a medical device which falls under a different set of regulations? Is it a biologic, which is yet another set of regulations?

Depending on which of these regulations this product chooses to market up against, then these levels of bioburden, or bacteria, or the levels of pyrogens are determined according to the regulation.

Initially, the company wanted to file this as a biologic. It didn't meet the criteria for it. They had to re-start their clinical trials and try to file under a cosmetic regulation. That cost them a lot of money; they had spent about $1 million with the batch that they had previously made and now had to spend another million dollars to make a second batch to file under a different regulation. Now they finally finished all of that and sent us the final product for testing.

It still didn't meet these criteria. It passed on endotoxins, but it didn't pass on bioburden. We separated out some of the data to determine whether there was an error and went through a full investigation – which again requires time and money – and found that there weren't any errors in the investigation or the laboratory testing. They had to go back and determine whether they were going to file another regulation – another million dollars in production for a small batch of this experimental cellular matrix.

That's an example of how if the decision isn't made in the beginning with a risk-based mindset, understanding justification, and doing good follow-through, it can be extremely costly. It could be that a million dollars is not a lot of money to a massive company, but it was to this particular company, which was just starting out with limited funds and limited venture backing.

Time was costly as well. It took six months to create a batch of cellular matrix. That was a problem. Was the company even going to have a viable product in two years? We all hope that we will have products that are going to help other people, save lives, and do good for the world, but sometimes these products don't work out. Knowing that sooner rather than later is better than spending a lot of time and money.

Here are some of questions you can ask yourself when you're trying to decide, as well as tools for creating the blueprint for your process and your product – big-picture questions:

- What are the things that should be considered along the way?
- How does quality fit in with manufacturing?
- When does quality need to be involved?
- What are the company goals?
- Does this fall into the correct set of regulations that you want to be under?

After those are answered, then we can get into details about what procedures need to be in place. These big-picture questions need to be answered first, so you have a good blueprint and a good plan to move forward, using your risk-based mindset with justifications and follow-through for your particular product. Check those out on the website and call us if you have any questions.

www.UltimateLabsInc.com

Printed in Great Britain
by Amazon

47245560R00027